UNDERSTANDING NATURAL FIBRE CONCRETE

Its Application as a Building Material

by

Barrie Evans

Practical Action Publishing Ltd
27a Albert Street, Rugby, CV21 2SG, Warwickshire, UK
www.practicalactionpublishing.org

© Intermediate Technology Publications 1986

First published 1986\Digitised 2013

ISBN 10: 0 94668 877 X
ISBN 13: 9780946688777
ISBN Library Ebook: 9781780443706
Book DOI: http://dx.doi.org/10.3362/9781780443706

All rights reserved. No part of this publication may be reprinted or reproduced or utilized in any form or by any electronic, mechanical, or other means, now known or hereafter invented, including photocopying and recording, or in any information storage or retrieval system, without the written permission of the publishers.

A catalogue record for this book is available from the British Library.

The authors, contributors and/or editors have asserted their rights under the Copyright Designs and Patents Act 1988 to be identified as authors of their respective contributions.

Since 1974, Practical Action Publishing has published and disseminated books and information in support of international development work throughout the world. Practical Action Publishing is a trading name of Practical Action Publishing Ltd (Company Reg. No. 1159018), the wholly owned publishing company of Practical Action. Practical Action Publishing trades only in support of its parent charity objectives and any profits are covenanted back to Practical Action (Charity Reg. No. 247257, Group VAT Registration No. 880 9924 76).

PREFACE

Many people have heard of natural fibre concrete technology and may be wondering if it is appropriate for them. This report is intended as a basic primer and provides an understanding of how the technology works and the current state of the art. It also indicates what distinguishes the technology from conventional mortars and reinforced concretes; what are the critical aspects of manufacture and installation; and what performance to expect of the building elements created. However the report is not a manual for making building elements, though a few sources of guidance are referred to.

This report was commissioned by the Building Materials Panel of Intermediate Technology Development Group and was undertaken by Barrie Evans, M.Sc. who is a Technical Editor of the Architects Journal. Natural fibre concrete is a technology felt to have potential in economically less developed countries, and perhaps in developed countries too. The material consists of concrete, usually just a mortar, reinforced with natural fibres, normally of vegetable origin. The material is most commonly found as roof sheets and tiles.

The report provides a basic understanding of this technology. It draws together knowledge from a wide range of research and development sources, including discussions with several experts in the field. In particular the author would like to thank: Hans-Erik Gram of SBI, Stockholm; John Parry of JPM Parry & Associates; Dr Narayan Swamy of Sheffield University and Malcolm Wilder of Brighton Polytechnic. The presentation of information and opinions should not necessarily be taken as the views of those individuals (or their organisations) who gave their help and advice so readily.

Michael Parkes
Manager
Mineral Industries Programme
ITDG
April 1986

CONTENTS

1. Introduction 1
2. An intermediate technology 4
3. Basics of strength 10
4. The potential of fibre reinforcement ... 17
5. Tensile strength of FC 20
6. Bending strength of FC 24
7. Toughness of FC 28
8. Durability of fibre and FC 32
9. Manufacture 35
10. Installation and use 39
11. Conclusion 42

References and notes 43

1. INTRODUCTION

Adequate shelter is a pressing need in many of the poorer parts of the world. As a developing technology, natural fibre concrete offers the possibility of making a variety of building elements. As an intermediate technology, it enables these building elements to be manufactured locally on a small scale at low cost.

Natural fibre concrete (FC) is made up of concrete (mortar) and natural fibres, such as sisal, jute or coir. It is formed into thin sheets or applied as a render (a thick surface coating). It is typically used 1 cm (½ inch) or less thick. So the building elements can be relatively lightweight per unit area. And they can have a relatively low cement content, helping to keep down costs.

Fig. 1 An example of natural fibre concrete roof sheets, on houses in Colombo, Sri Lanka.

The most critical element in the building of shelter is usually roofing. To date, this is where natural fibre concrete has found its main application, in the forms of corrugated roofing sheets and more recently, pantiles. Other applications include wall cladding sheets, culverts, road surfaces, and protective renders to sun-dried brick or mud walls.

STATE OF THE ART

Applications of fibre concrete are still in their infancy. This is true both for natural fibres and for more industrialized versions of the technology using fibres such as glass. Only asbestos cement corrugated sheeting manufacture is established so far as a reliable fibre concrete technology, though not as an intermediate technolgy.

As with any developing technology, there has been both progress and problems. Roofing sheets have been installed successfully in many countries around the world, although there have also been problems, primarily in manufacturing/installation quality control and in fibre durability.

Manufacturing/installation quality control

Natural fibre concrete is not a technology like the sewing machine which can be picked up and used almost without instruction. A large amount of knowledge has been accumulated on manufacture and installation, especially of roofing sheets and tiles. Any prospective manufacturer or installer needs to share this knowledge in order to make building elements reliably at reasonably low cost and to introduce them successfully. Those who have tried to the copy the technology from brief magazine articles, reports or word of mouth descriptions have usually been unsuccessful.

The addition of fibres to a mortar mix does offer several advantages over unreinforced mortar. It facilitates manufacturing and installation quality control: fibre aids cohesiveness of the wet mix, helping to hold it together while it is being trowelled and shaped: fibre also helps limit drying shrinkage cracking which occurs as the wet material dries out; and fibre helps improve some of the early strength properties of the material, so improving the chances of manufactured building components getting to site and being installed undamaged.

Fibre durability

The other area of concern is fibre durability. Natural fibres may decay in concrete from within months to a few years of manufacture. No convenient, cost effective method has yet been established to significantly extend natural fibre life. Inevitably, most of the extra strength properties that fibre helps create initially are lost as the fibre decays.

However, building elements can retain sufficient strength to be serviceable for a wide variety of applications. For example it is assumed that the natural fibres have by now decayed in most of the natural fibre concrete roofing sheets and tiles now installed around the world. Provided these were well made and installed, they generally continue to be effective.

Where there have been failures, it has only occasionally been because their use required them to be stronger than they became following fibre decay. Usually, any failures are attributable to poor manufacturing or installation quality control.

The technology must be transferred with care and awareness of its scope and limitations. If this is done it will be successful.

2. AN INTERMEDIATE TECHNOLOGY

At its simplest, fibre concrete is a mix of cement, sand, water and fibres. As an intermediate technology, naturally occurring fibres are used. A wide range of fibres has been investigated. The commonest successfully used are probably sisal, henequen, coir, jute and manilla. Their common use though is more to do with their wide availability than any exceptional suitability in strength or other properties. Many other fibres may also be used, from flax fibre to banana bark.

Fig. 2 Detail of timber-frame construction for supporting a fibre concrete roof.

MANUFACTURE

Manufacture is in principle straightforward, following either of two methods which use short, chopped fibres (12-50 mm, ½-2 inches), or long fibres. The short fibre method is used for renders and most building components. The long fibre method is also used for some building components, though it is less well developed and thus potentially less reliable as a manufacturing process. (There are other relative advantages and disadvantages of the methods too, which we will come to later.) Roofing components made by the short fibre method are believed to be in use in as many as 25 countries, whereas the long fibre method is in use in only two.

Short fibre method: All the ingredients are carefully measured out and then mixed together - sand, cement, water and chopped fibre. The mix is trowelled onto a flat surface - figure 3.

Fig. 3   An even thickness of material is created by trowelling within a shallow timber frame. This example is a step in roof sheet manufacture by the short fibre method.

Long fibre method: The concrete - sand, cement, water - is mixed. A thin layer, say 3mm (approx. 1/8 inch) thick is trowelled onto a flat surface. Fibres are laid on top, aligned with the long direction in the case of roof sheets and tiles, and worked into the concrete with a roller - figure 4. This may be repeated. Typically two or three layers of concrete sandwich one or two layers of fibre. The first one or two concrete layers are around 2mm, the last one is thicker to make up the full sheet depth of say 8-10mm (approx 3/8-½ inch). Thus the long fibres are sandwiched in the bottom half of the sheet - figure 5.

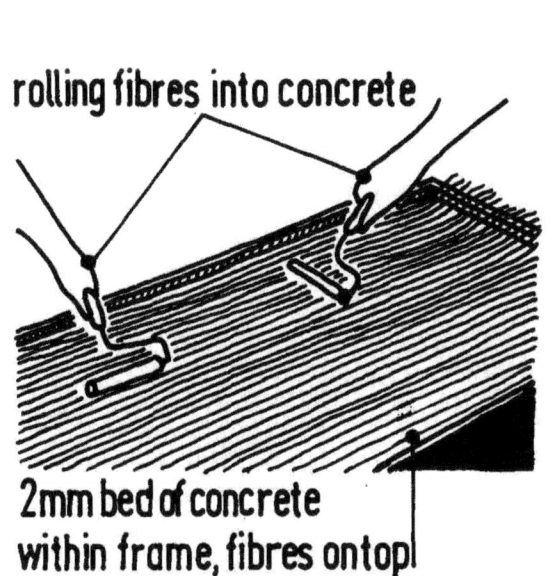

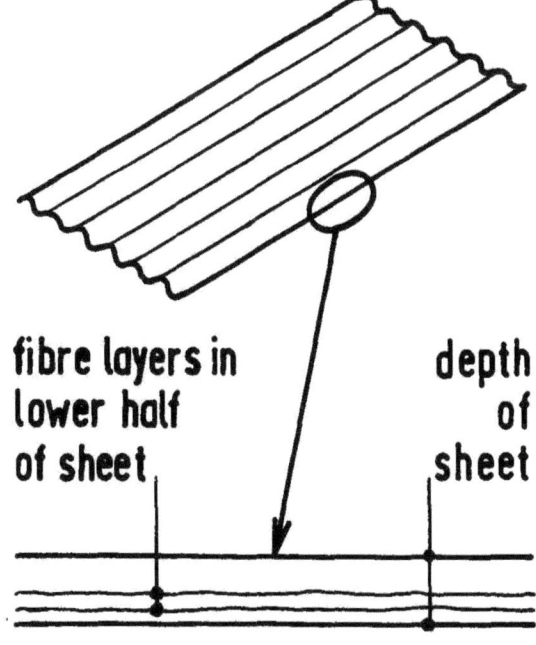

Fig. 4   Using rollers to work long fibres into the first thin layer of mortar.

Fig. 5   Section (slice) through a roof sheet illustrating the way long fibres are sandwiched in the lower part of the sheet.

For both methods, once trowelled flat the mix is tamped to consolidate it and drive out air. (Tamping here refers to repeated hitting of the edge of the mould or surface supporting the wet mix.) Recently a more elaborate but more efficient tamping process has been introduced for pantiles, making them on a small motor-driven vibrating table - figure 6. In the long fibre method, the fibre is intended to reinforce the concrete (as in steel reinforcement), whereas in the short fibre method, no actual reinforcement is intended.

Fig. 6 Trowelling fibre concrete on a motorized vibrating table in producing a pantile.

Building components are often shaped rather than flat, such as corrugated sheets or pantiles. Shaping is usually achieved by lowering the wet mix onto an appropriate mould - figure 7.

Fig. 7 A later stage in making a shorter (1m) roof sheet using either long or short fibre method.

There is more to manufacturing than this outline has explained (see also chapter 9). Evidently though, manufacture can be an intermediate technology process - small-scale, local, relatively simple and cheap. But how does it compare with other existing alternatives? Is the choice of natural fibres an advantage?

ALTERNATIVE FIBRES

Several non-natural fibres are available, especially in industrialized countries. Short fibres are generally used because of the relative ease of manufacture and the structural properties sought in the finished products. Common among these fibres are polypropylene, steel, glass and asbestos.

Polypropylene fibres

These can be supplied as short fibres or as a tape or string for chopping. The fibre is roughened on the surface in order to bond well with the concrete. Polypropylene has a lower health risk than asbestos. Polypropylene can be used at fibre contents well below one per cent by weight of the concrete. (This compares with about 15 per cent for asbestos fibre.) Steel and glass can also be used at such low fibre contents - one to two per cent - but are more expensive than polypropylene. Polypropylene is more resistant to decay from the alkaline environment within concrete than are natural fibres. So polypropylene fibres could be used as an alternative to natural fibres. They will almost always be a more expensive import except when found as discarded flour sacks. Natural fibres are usually locally available, very cheap and renewable.

Steel fibres

Steel usually begins as a tape for chopping although some prepared scrap can be used. Steel is not very appropriate for thin building elements because it is sharp, so tending to damage hands, and is stiff, so tending to stick out on the element's surface after manufacture. (Steel fibre concrete should not be confused with ferrocement which is a steel mesh framework that is rendered over with mortar.)

Glass fibre

Manufacture of glass fibres is a complex technology, so glass fibre would usually have to be imported. These fibres are not cheap, though the percentage fibre content is of course small. Special alkali-resistant glass is used for the fibres used in this application. More restrictive is the licensing arrangement. In many cases fibres can only be manufactured or incorporated into building components under factory conditions monitored by the licenser. This is clearly inappropriate for small-scale intermediate technology manufacture, often in remote locations.

Asbestos fibres

Although asbestos is a natural product, it is not included as such in this report which deals exclusively with natural fibres of vegetable origin. Setting up an asbestos cement sheet manufacturing plant could well cost £1 million. On top of that are the serious health risks of mining and working with asbestos. While asbestos cement roofing sheets are made in several less economically developed countries, there is no known way of scaling down manufacture to become a small-scale, intermediate technology. (For further information on these fibre technologies see reference 1.)

Asbestos cement and galvanised iron sheets

For roofing applications, asbestos cement and galvanised iron are the most widespread of non-traditional roofing products in economically less developed countries. They are well established. It is with these (and thatch) that any roofing application of natural fibre concrete usually competes. The material can be expected to be competitive on a mix of price and performance. But the market may have different priorities, for example choosing lower quality galvanised iron if it is cheaper, or the price advantages of natural fibre concrete may be too marginal to make it worth the risk of setting up manufacture against established alternatives.

CEMENT AVAILABILITY

One other factor that may affect natural fibre concrete's viability as an intermediate technology is the availability or cost of cement. All other ingredients - sand, water, fibre - should be easily and cheaply available. Difficulties in obtaining adequate cement at the right price would on occasion make the technology less appropriate or inappropriate altogether.

Cement substitutes

Various so-called cement substitutes might be included in the mix^2. In fact these would normally be partial cement substitutes. Some cement would still be needed.

A partial substitution of cement with a pozzolanic material would have the advantage of reducing alkalinity of the cement matrix and therefore reducing the rate of fibre decay. Research and development is needed before any substitutes could become a dependable part of natural fibre concrete technology.

3. BASICS OF STRENGTH

First, a definition. Having discussed alternatives to natural fibres the remainder of this report focuses only on natural fibre concrete. For this we shall use the abbreviation FC (Fibre Concrete), taking as understood in this report we are referring to 'natural (vegetable)' FC.

Those readers familiar with building will have noted that the concrete in FC is in fact mortar, a mix of sand, water and cement. True. But the term FRM is rarely used.

On other occasions reference is made to fibre <u>cement</u>. This label is appropriate where the mix is fibre, water and cement, omitting the sand. For example asbestos cement.

The initials FCR (Fibre Concrete Roofing) are also used, for example by the study group referred to in reference 3. But roofing is only one of the applications of FC to which this report applies so this is not used either.

STRENGTH PROPERTIES

This and subsequent chapters discuss various key strength properties and the prospects of achieving them in the typical conditions of intermediate technology manufacture. This chapter sets out basic structural concepts. Anyone with at least a rudimentary knowledge of these concepts need only perhaps check the last two items - 'Elastic modulus' and 'Toughness' - before moving on to the next chapter.

STRESS

When a <u>load</u> is applied to an <u>area</u> of a structural element it is said to be under <u>stress</u>, as represented by load per unit area.

$$\text{STRESS} = \frac{\text{LOAD}}{\text{AREA}}$$

TENSION

If we pull an element it is in tension. By applying this tensile load we generate a <u>tensile stress</u> within the element - figure 8.

COMPRESSION

If we press an element it is in compression. By applying this compressive load we generate a <u>compressive stress</u> within the element - figure 9.

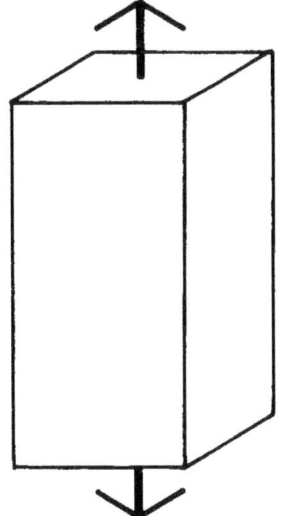

Fig. 8 A structural element in tension (being pulled apart). The element is subject to tensile stress.

tension

Fig. 9 A structural element in compression (being pressed together). The element is subject to compressive stress.

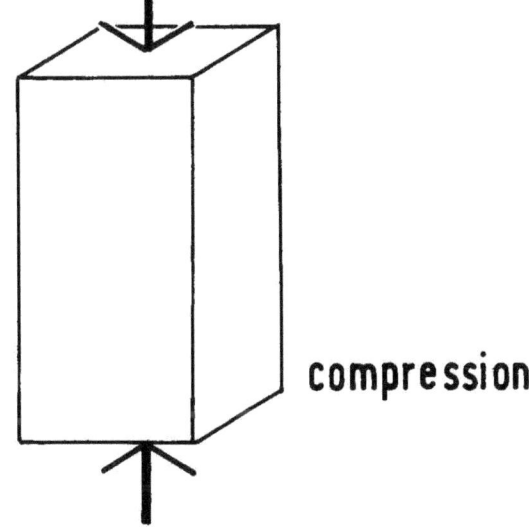

compression

BENDING (also known as FLEXURE)

If we support a building element say at its ends and apply a load to one face, it bends. This is illustrated for a simple beam in figures 10ab. In practise this bending action may be too small to be seen, as for example in the case of a steel or concrete bridge. It occurs nonetheless. As figure 10b shows in exaggerated form, the top of the beam is pressed together - it is subject to compressive stress. The bottom is stretched - it is subject to tensile stress.

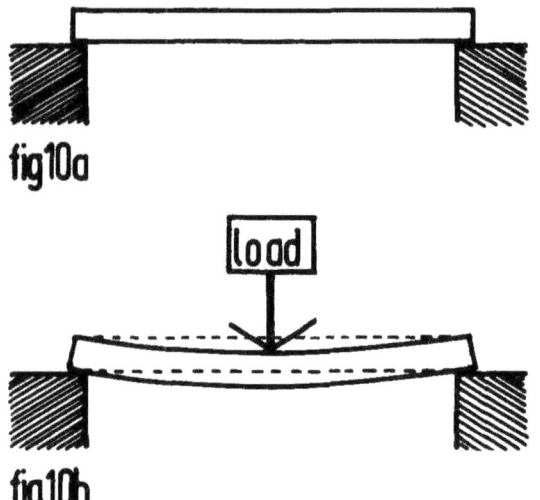

Fig. 10 For a beam bridging between supports, as in 10a, an applied load will cause it to deflect. Bending stresses are set up within the beam. The deflection is made obvious here, 10b.

It is not of course just the edges that are stressed. There is a variation of stress through the whole element ranging from a maximum at the edges. The graph - figure 11 - illustrates this stress variation for compression at the top of a beam and tension at the bottom. The location of changeover between compression and tension is known as the neutral axis, shown both in figure 11 and in figure 12.

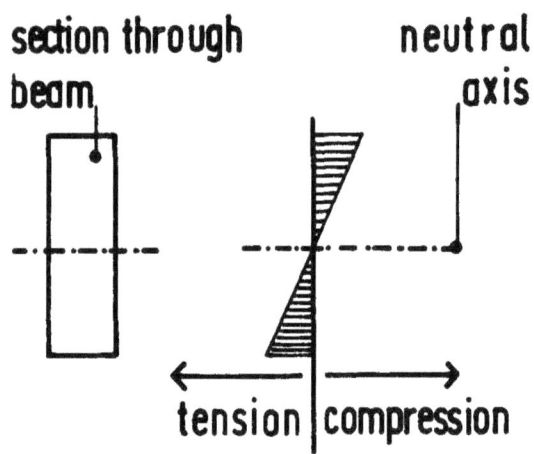

Fig. 11 A beam in bending, like that in 10b, is compressed at the top and stretched (in tension) at the bottom. Figure 11 shows a section across the beam - left of diagram - and to the right a graph of the distribution of stresses induced by bending. At the top surface is a maximum of compression. At the bottom is a maximum of tension. The centre is called the neutral axis, the level at which there is a changeover between compressive stress and tensile stresses. At the neutral axis there is no stress.

Fig. 12 This time looking along rather than across the beam, (as did figure 11). The neutral axis runs the length of the beam at the same level.

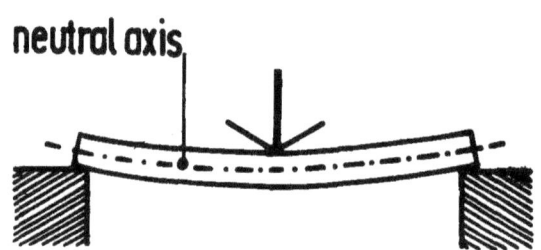

FC in bending

The neat stress distribution graph - figure 11 - is for a beam of uniform cross section made of a homogeneous material that has equal strength in compression and tension (hence the neutral axis being located at the midpoint). The general principle of compression on one side of a neutral axis, tension on the other, applies to any structural element in bending. But as we shall see later, FC is not a homogeneous material, it is a _composite_ of mortar and fibre. The mortar itself is stronger in compression than in tension. The fibres and the mortar work together in different ways depending on the amount of stress and whether it is in compressive or tensile. The neutral axis thus varies in position. So the pattern of bending for a real FC composite is more complex than shown here for a simple beam.

Bending stress capacity of a beam increases with the depth of the beam. Thus if we have a beam of rectangular cross section loaded as in figure 10b, the beam can take a greater load if the long edge is vertical. (This assumes that the beam is not so slender that it buckles in the compression zone.)

We can transfer this understanding of the benefit of depth in bending to thin FC elements. An FC sheet would have little bending strength if it were flat. But corrugating it like a corrugated roof sheet or pantile introduces depth and greatly increases its bending strength capacity for the same amount of FC material.

STRAIN

As a building element is loaded it deforms, again often unseen by the eye. For a structural element in tension or compression the fractional longitudinal extension or shortening is called the _strain_. This deformation is conventionally measured as a percentage of the original length. While strain is most straightforwardly measured for tension or compression, strain measurements can be made for bending, say on the outside surfaces of the element.

ELASTICITY

As for many structural concepts, elasticity has a more precise meaning than in everyday speech. Any solid material, whether rubber band, glass or mortar exhibits some elastic behaviour. As a load is applied, deformation (shape change) occurs. If when the load is removed the material recovers its original shape, the temporary shape change is called _elastic deformation_.

Elastic deformation is expected and designed for in structural elements. Their elastic behaviour is usually their prime structurally useful behaviour. Normally this is <u>linear elastic</u>, that is, the amount of strain is proportional to the amount of stress. Twice the stress, twice the strain. If you draw a graph of stress against strain you get a straight line. Where this elastic behaviour stops is called the elastic limit - figure 13.

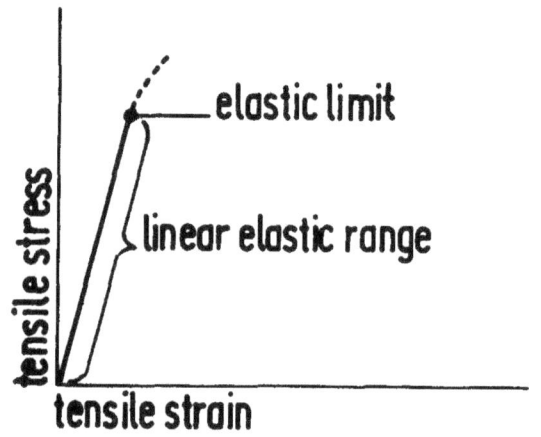

Fig. 13 Graph of stress against strain in tension.

PLASTICITY

Under an applied load a material deforms but may not recover some or all of its original shape when the load is removed. The resulting shape change is called <u>plastic deformation</u>. Structural materials often deform elastically, reach their elastic limit, then begin to deform plastically. This latter behaviour would normally be elasto-plastic, that is, it would combine some temporary elastic deformation with some permanent plastic deformation. For many materials elastic behaviour and elasto-plastic behaviour are distinct phases, as illustrated in figure 14.

Fig. 14 A common but by no means universal relationship between stress and strain beyond the elastic limit. It is common for the elastic and elasto-plastic deformations to be clearly distinguishable forms of structural behaviour.

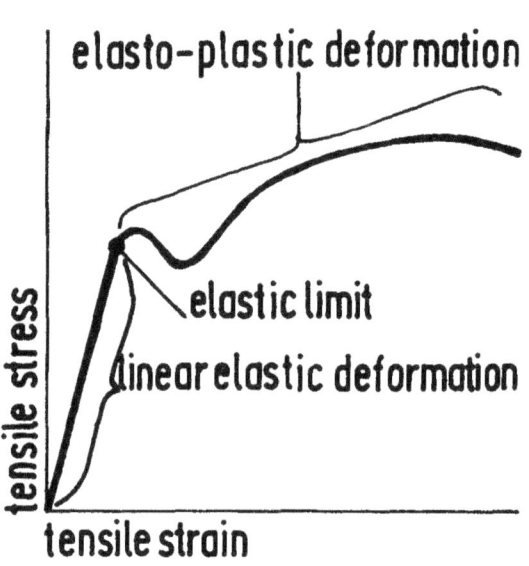

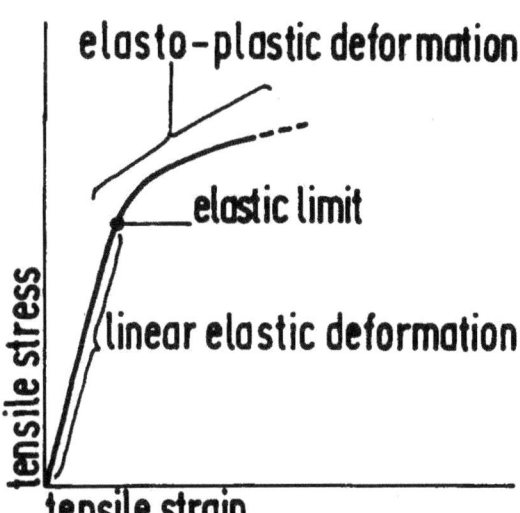

Fig. 15 Indistinct transition between linear elastic and elasto-plastic structural behaviour.

For some materials the elastic limit is ill-defined and elastic behaviour shades into elasto-plastic - figure 15. For some brittle materials such as mortar there is very little elasto-plastic deformation before the material cracks and collapses - figure 16.

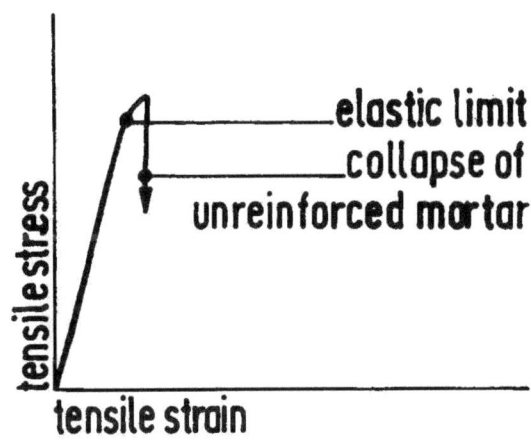

Fig. 16 After very little elasto-plastic deformation, unreinforced mortar collapses (and so is unable to take any stress at all).

ELASTIC (YOUNG'S) MODULUS

'Linear elastic' behaviour was mentioned above. This approximately constant relationship between stress and strain can be expressed as a mathematical constant called the Elastic Modulus, (also known as Young's modulus).

$$\text{Elastic Modulus} = \frac{\text{STRESS}}{\text{STRAIN}}$$

What sort of property is an elastic modulus? Materials that exhibit little strain for a given stress are called high modulus materials. They are stiff. Those which exhibit considerable strain for the same stress, extending or compressing a relatively large amount, are called low modulus materials. They are stretchy.

The terms 'high' and 'low' modulus are of course relative not absolutes. To call a modulus high or low depends on the application being considered.

TOUGHNESS

'Toughness' too has a precise structural meaning, if somewhat difficult to grasp as a concept.

Tough materials have a useful resistance to deformation. This somewhat vague statement can be clarified by contrast with other materials. Toughness can be thought of as a condition intermediate between brittleness and softness. Brittleness is the tendency to fracture with little deformation, like glass. Softness is the tendency to deform easily, like a pillow.

More technically, we can talk about the fracture toughness of a material. Once beyond the elastic limit brittle materials (like mortar) fail by major cracking - figure 16. Tougher materials have a greater capacity to inhibit the propagation of these disastrous cracks by absorbing more of the breaking load (fracture energy) through more localised multiple microcracking. Forming these cracks absorbs energy.

The energy level at the tip of a propogating crack can be say 100 times that of the average stress in a material. So crack inhibition and crack control can be crucial to the continuing serviceability of a structural element.

IMPACT RESISTANCE

Impact resistance is not a fundamental property of materials, though it is allied in some ways to toughness. In absorbing impacts the material absorbs energy. Impact resistance is most easily explained by the way it is measured. A standard weight is dropped a standard distance onto a sample of the material. The degree of indentation is the measure of impact resistance. This can be converted to a figure for energy absorbed.

4. THE POTENTIAL OF FIBRE REINFORCEMENT

Unreinforced concrete is not an ideal material for making building products, especially thin ones. Adding fibres provides the potential for improving manufacturing quality control - notably increasing cohesiveness of the wet mix when shaped and limiting cracking due to drying shrinkage. And fibre addition may also improve key strength properties - bending strength and toughness.

COHESIVENESS OF THE WET MIX

In the wet state an unreinforced concrete mix is not very cohesive, that is, it does not hold together very well. In industrialized manufacture thin unreinforced mortar as used for roof sheets and tiles is helped to hold together by heavy pressing against a mould. As we saw in chapter 2, intermediate technology FC manufacture has no pressing process. Especially where shaped building products such as corrugated sheets are being made by being lowered onto a mould - as in figure 7 - wet unreinforced mortar is likely to crack. This occurs notably along ridges such as the top of rolls of corrugated roof sheets where stretching is greatest - figure 17.

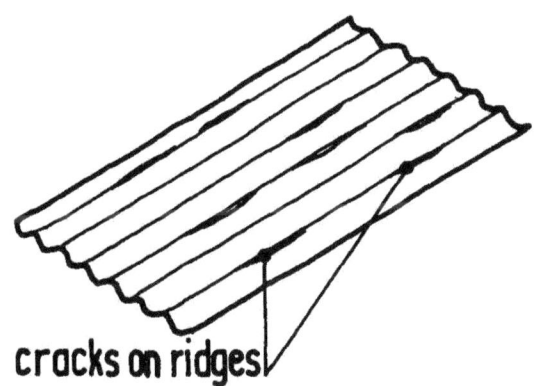

Fig. 17 Risk of cracking of wet mix is greatest where curvature and hence stretching is greatest, in this case along the top of ridges where the wet mix is formed into a corrugated sheet.

With the addition of chopped fibre, wet FC can with practice be more reliably transferred to the shaped mould. FC is more cohesive than an unreinforced mortar, it exhibits a useful plasticity in the wet state. Fibres help hold the material together.

Long fibres appear less suitable for increasing cohesiveness of the mix than short fibres since their effect is primarily in the direction of fibre alignment. In particular, for roof sheets and tiles, long fibres lie predominantly along the ridges offering less help in binding the wet mix across the ridges to counter the danger of cracking illustrated in figure 17. Additional fibres could be laid across the ridges to help counter this problem.

DRYING SHRINKAGE CRACKING

Concrete shrinks as it dries. There is more water in a mix than is needed for the chemical reactions of setting. How much more water is added is a compromise. The more water the more workable the mix is, that is, the easier it is to handle in the wet state. For example it is easier to spread with a trowel.

However, an increase in the ratio of water to cement increases the risk of flaws from drying shrinkage cracking.

Paradoxically adding fibres to mortar reduces workability of a mix and so more water is usually added. However the benefits fibres bring in controlling drying shrinkage cracking far outweigh the increased risk of cracking arising from the additional water. There is a net benefit in shrinkage crack control. An increase in water/cement ratio also results in greater permeability and a decrease in strength. With a relatively small increase in the ratio, this is unlikely to be significant.

The ancient practice of mixing straw into bricks or cow hair into plasters exemplifies this same principle of drying shrinkage crack control. The significant part of the Exodus story of Pharaoh denying the Israelites straw for brick making is:

> 'So the people were scattered throughout
> all the land of Egypt to gather stubble
> instead of straw'.

They had to find something. They could not make sun-dried bricks reliably without it.

As with cohesiveness, so with drying shrinkage cracking. It may be that short randomly oriented fibres would be more effective than long fibres which lie in a single direction. The compensation of long fibres is that they form a stronger bond with mortar than short fibres. There is no systematic data on these differences between fibres for cohesiveness or drying shrinkage crack control.

BENDING STRENGTH

Unreinforced concrete is much stronger in compression than in tension. Bending an unreinforced beam for example would lead to failure by cracking in the tensile zone - figure 18. In conventional reinforced concrete, steel rods are put into the tensile zone as tensile reinforcement to compensate for concrete's own lack of tensile strength. Thereby, tensile and compressive strengths are brought into balance.

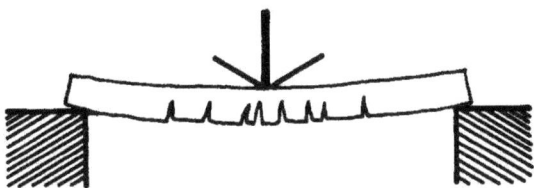

Fig. 18 Possible cracking pattern of unreinforced concrete beam. Cracking occurs first in the bottom of the beam, which is in tension, because unreinforced concrete is much weaker in tension than in compression.

The potential is to use natural fibres similarly to steel reinforcing rods. In the case of long fibres these are aligned appropriately for bending, concentrated in the tensile zone and being long bond well with the mortar. Short fibres are randomly oriented and uniformly distributed throughout tensile and compressive zones, and being short bond to mortar less well. Long fibres are much more effective than short fibres in acting as tensile reinforcement and so contributing to bending strength of the FC composite. The nature of the tensile and bending processes are discussed in the next two chapters.

TOUGHNESS

In describing toughness in the last chapter, fracture toughness was mentioned. Tougher materials are more resistant to the propagation of major cracks and have greater potential to absorb fracture energy through the process of multiple microcracking. Fibre can help in this, holding FC together, inhibiting major crack growth. Fibre pull-out from the mortar during microcracking is itself a significant energy absorbing process. The presence of fibres should make significant improvements in toughness. These factors are discussed in chapter 7.

5. TENSILE STRENGTH OF FC

Having indicated that natural fibres offer the potential to increase strength properties of mortar, we look at the mechanisms by which this occurs. We look in this chapter at tensile strength as a first step in understanding the more useful but complex property of FC - bending strength (covered in the next chapter).

Figure 19 is a highly idealized stress/strain diagram for FC in tension. It provides a central focus for setting out the tension story of FC. Each region of the graph is discussed in turn (For a fuller, more technical discussion of the processes see reference 1.)

THE 'A-B' REGION

The region marked 'A-B' on figure 19 is the elastic behaviour of FC - like figure 13. Point 'B' is the elastic limit. We would like to increase the tensile strength of FC beyond that of mortar alone, which is relatively weak in tension, that is to raise point 'B'.

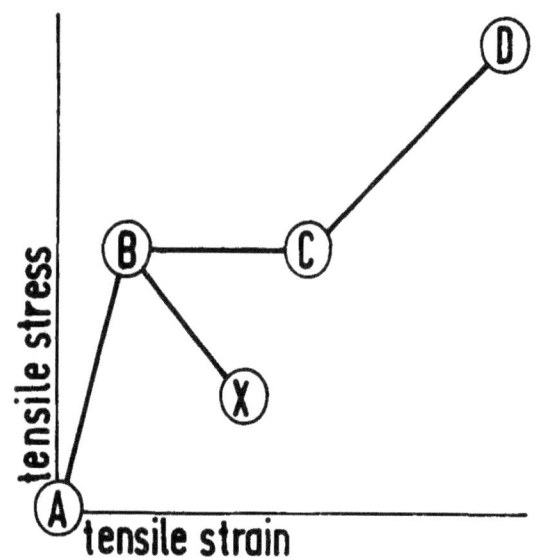

Fig. 19 Idealized graph of stress against strain for fibre concrete in tension. The region 'A-B' is linear elastic, with point 'B' the elastic limit. The region 'B-C-D' is the elasto-plastic behaviour of FC reinforced with long fibres. The region 'B-X' is the elasto-plastic behaviour of FC incorporating short fibres. (see reference 1)

Unfortunately no significant benefit will be achieved by fibre addition. (Some laboratory tests have found small improvements, others none, others still have found small reductions.) The theory of mixtures says that stress induced in either part of the composite - mortar and fibre - will be proportional to their respective elastic moduli. The part with the higher modulus will strain less for a given stress and take on proportionally more of the load. There are two problems in getting natural fibres to do this in FC:

- The elastic modulus of mortar is around 20-25 GN/n^2, (ignore these units, GN/m^2; it is the relative size of the numbers that matters). The elastic moduli of common natural fibres are similar to that mortar[5] - jute 26-32 GN/m^2, sisal 13-26 GN/m^2, coir 19-26 GN/m^2. Broadly speaking, up to the elastic limit natural fibres behave in tension like mortar. They bring no significant improvement to the composite.

- Ideally then we would look for a high modulus fibre similar say to steel (200 GN/m^2) or asbestos (165 GN/m^2). But even such high modulus fibre would be of limited help. As mentioned above, stress borne for a given strain is proportional to the elastic modulus. If a natural fibre had an elastic modulus 10 times that of mortar it would take 10 times the stress for a given strain. But stress is load per unit area. In any cross section of FC there is a relatively small area of fibre, at most a few per cent. These high modulus fibres would be heavily stressed individually, but taken together they would contribute only a small fraction to load bearing capacity of the FC overall[6].

To summarize, increasing the tensile strength of mortar significantly would require high modulus natural fibres in quantity. Common natural fibres do not have a high modulus compared with mortar. And manufacturing practicalities limit the fibre content to a few per cent. The design value for elastic tensile strength of FC is the same as that of the mortar.

Note too that this mortar strength is somewhat less than mortar would otherwise have, because extra water is usually added to the mix along with fibres to keep the mix workable. This increase in water/cement ratio reduces elastic tensile strength of mortar by a small but noticeable amount.

THE 'B-C' REGION

The region 'B-C', and subsequently 'C-D' represent the elasto-plastic tensile behaviour of FC with long fibre reinforcement.

While the elastic moduli of mortar and natural fibres are similar, their ultimate (breaking) tensile strengths are not. Natural fibres are usually much stronger. The ultimate tensile strength of mortar is around 5-8MN/m^2 (ignore the units again). By contrast figures for fibres[5] are for example coir 120-200MN/m^2, jute 250-300MN/m^2, sisal 280-568MN/m^2. Thus, once the elastic limit of mortar is passed with its brittle mortar failure it is possible to keep adding load to the fibres (provided there is sufficient fibre content).

In the region 'B-C' the mortar gradually cracks. There is high stress at the crack tips as they spread. Fibres span these cracks and inhibit their propagation to some extent. Rather than a few disastrous major cracks, multiple small-scale cracking occurs, typically at less than 5mm (approx 1/4 inch) spacing and virtually invisible to the untrained eye. The FC effectively becomes a set of mortar blocks held together by fibres. The tensile load is gradually transferred to the fibres.

For this theoretical behaviour to occur, adequate bond of fibre to mortar and adequate fibre content is required. Suitably strong long fibres can provide adequate bond of fibre to mortar. It is possible to include natural long fibres in adequate quantity and achieve good enough bond for the load to be transferred to the fibres when the mortar fails in tension.

THE 'C-D' REGION

In this region, which applies to long fibres, the fibres are increasingly loaded. The FC fails both by fibres themselves breaking in tension and by the fibres pulling out of the mortar.

However, not all this extra strength over unreinforced mortar may be particularly useful in pratice. The form of material represented by regions 'B-C-D' changes once only; they are irreversible. In region 'C-D', before reaching 'D', cracks may open up to such a size either that the FC building element is no longer weatherproof or that the fibres are exposed to air and so to the risks of biological deterioration.

THE 'B-X' REGION

This region represents tensile behaviour of FC incorporating short fibres, beyond the elastic limit. Short fibres do not help increase tensile strength beyond that of the mortar's elastic limit. At least three contributory reasons can be cited for this:

- If more than one to two per cent of short fibre is added to mortar in making FC it tends to collect in balls rather than being evenly and separately distributed. So around one to two per cent is a practical upper limit. This is less than can be achieved with long fibre. And given that short fibres are randomly distributed rather than being concentrated in the critical tensile zone, the effective concentration of short fibres is even less compared with long ones.

- The strength of bond of fibre to mortar is limited among other things by the fibre length. Increasing the length of 'short' fibres to say 50-75 mm (2-3 inches) or more does not help much because fibres bend and twist in mixing. Their effective length is only similar to short fibres 15-25mm (approx ½-1 inch) or less[7]. Bond strength is significantly less than for long fibres.

- While long fibres are aligned with the direction of bending stress, short fibres are randomly oriented. Their effectiveness in tension in the single direction of tensile stresses averages around 1/5 - 1/6 of that of long fibres[8].

For these reasons - fibre content, bond, orientation - short natural fibres do not add to the tensile strength of mortar. For design purposes the ultimate (breaking) tensile strength of short fibre FC is the tensile strength of the mortar, represented by point 'B'. And as mentioned earlier in this chapter for region 'A-B', the extra water added with the fibre to keep the mix workable reduces this tensile strength of mortar to some extent compared with a mortar mix without fibre and extra water.

6. BENDING STRENGTH OF FC

Bending strength is usually more important than tensile strength for building products. Looking at tensile behaviour first has been useful because that is what in essence goes on in the tensile zone of a building element in bending. The tensile zone is the complex one. Fibres have no significant compressive strength and so are irrelevant to the compressive zone.

However bending is not as clear-cut as tensile behaviour for several reasons:

- As we showed in figure 11, stress varies across an element in bending from extremes at the outside surface to no direct stress induced by bending at the neutral axis. If an FC element in bending fails in the tensile zone it will do so progressively from the outside edge where stress is highest inwards, rather than the whole tensile zone being uniformly stressed and all failing at once.

- The picture is further complicated by the fact, as we saw in regions 'B-C-D' of figure 19 that FC can sustain considerable extension (strain) in tension before ultimate failure. However during bending, the compressive zone of the FC cannot sustain a comparable amount of shortening (strain). There is a tendency for crushing to occur in the compressive zone and the strain in the tensile zone to be less than that suggested by figure 19 for pure tension.

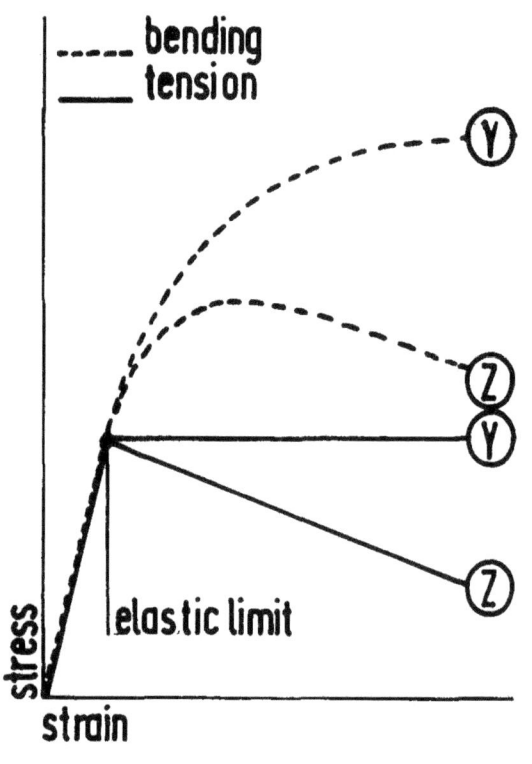

Fig. 20 Theoretical stress/strain graph for two samples of FC - 'Y' and 'Z'. In tension or bending there is linear elastic behaviour. But beyond the elastic limit there is a much more gradual transition into elasto-plastic behaviour for samples, if subjected to bending stresses than if under tensile stress. (For more explanation than is offered in the text, see reference 1, page 45.)

One implication of this mutual adjustment of mortar and fibre is that failure does tend to be less sudden than for mortar alone or for FC in tension. Theoretical curves in figure 20 for example compare idealized tensile behaviour with idealized bending for some generalized FC materials. Figure 21 is much closer to reality, a generalization of stress/strain curves for samples of natural fibre FC tested in the laboratory.

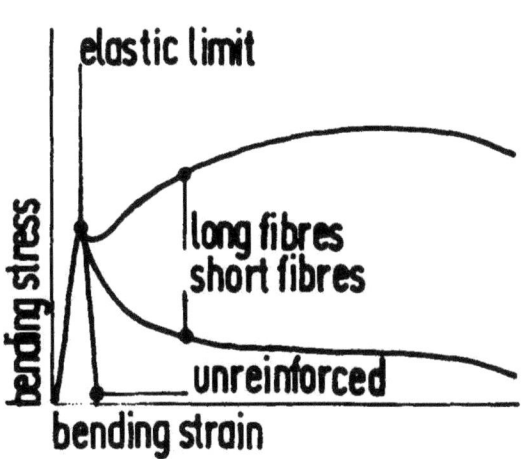

Fig. 21 Much nearer to the way FC behaves on site in bending than the previous theoretical curves in figures 19 and 20, (though still not a fully realistic record of stress/strain behaviour of a FC product in use).

SHORT FIBRE FC

Since we saw that short fibres add nothing much to tensile strength it is not surprising that they add nothing much either to bending strength. Failure is less sudden than in tension. For design purposes, the bending strength of short fibre FC is the elastic limit of the mortar alone, as shown in figure 21. As for tension, the extra water that tends to be added with fibre will reduce the elastic bending limit below that of a mortar without fibre or extra water.

LONG FIBRE FC

Figure 21 indicates that with adequate fibre strength, volume and bond, long, aligned fibres can increase ultimate (maximum) bending strength of FC beyond that of mortar alone. Some laboratory tests[9] for example have shown ultimate bending strengths for sisal reinforced FC of the order of three times that of mortar - figure 22.

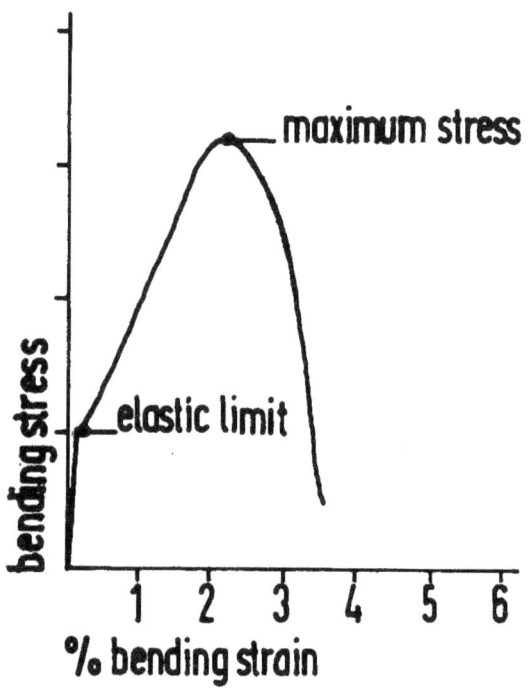

Fig. 22 Example of stress/strain behaviour in a laboratory test of a FC sample with long sisal reinforcement. Given that the unreinforced mortar would fail close to the elastic limit, reinforcing has produced around a threefold increase in bending strength (while fibres last). Note too though that the maximum stress occurs around 2 per cent strain, which can only be achieved with considerable cracking of the material. This elasto-plastic deformation corresponds to region 'B-C-D' in figure 19. (Test performed by Gram - see reference 9, page 81.)

The earlier note of caution needs repeating - that not all achievable strength is necessarily useful strength. The changes, especially cracking, are largely irreversilbe. And the high ultimate bending strength typically occur at high strains around two or three per cent. Approaching such strains, crack sizes may render the building element unserviceable even though it has not yet failed in bending.

Being able to achieve improved bending strength with long fibre reinforecment is an encouraging finding. If the problems of natural fibre's limited durability in concrete can be solved (see chapter 8), a range of interesting structural possibilities would open up. Not only stronger elements like roof coverings but primary structural elements such as hollow beams and flooring units - figure 23.

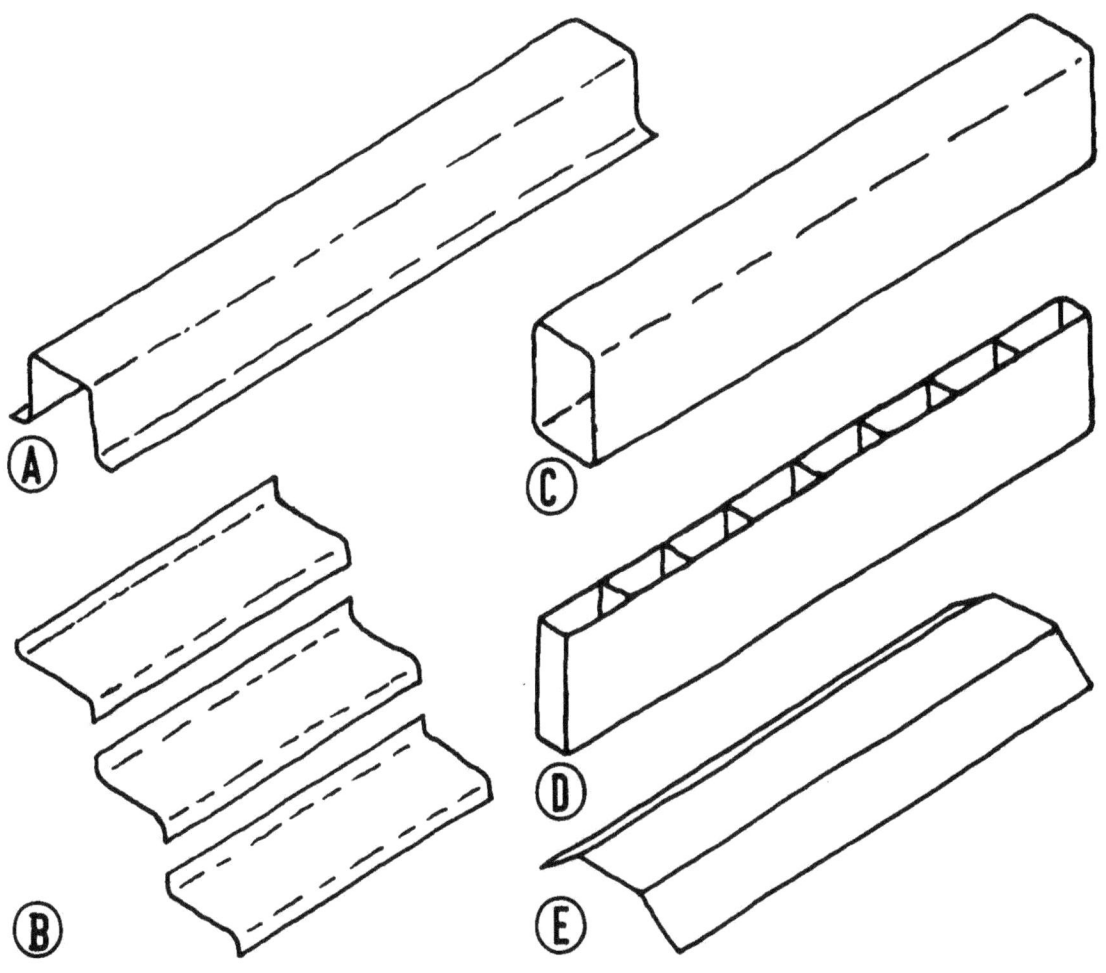

Fig. 23 If natural fibres could be made durable, then a variety of semi-structural possibilities suggest themselves. These (untested) options include beams and other means of bridging - A,C,D,E - and stair treads - B - for building-in between two parallel walls.

7. TOUGHNESS OF FC

In the case of bending just discussed, it is obvious that it would be useful to have the improvements in bending strength that fibre reinforcement helps create. Greater bending stresses could be supported. Thus for example a given building element could take higher imposed loads, could span further between supports, or could be made thinner. This latter in turn would save materials, most usefully the most expensive ingredient - cement. And in the case of roofing, a lighter weight of FC roof covering would reduce the weight supported by the roof structure so that a lighter roof structure might be possible. (Weight may not be the most critical load on the roof structure). There is a potential to use such strength increases in FC in a controlled way, by design, if the fibre durability problems can be solved. A wide range of groups have looked at bending strength in a systematic way (ref. 5, 9, 10, 11, 12, 13, 14).

The point of discussing aspects of bending strength in this section is to point out a situation that we regard as 'normal' with respect to strength properties. And thus to contrast it with the situation for toughness.

The most obvious difference is that we cannot design for conditions requiring toughness as a normal part of structural design because we do not know what the loads to be supported are with any precision. So while we have some indicative data on toughness that show one material as tougher than another, we cannot relate this directly to service conditions. And even the toughness data for FC are open to question. The most we can do in this chapter is to discuss the likely advantages of improving toughness in FC building elements.

BENEFITS OF IMPROVED TOUGHNESS

Building elements have to sustain both planned loading in use, e.g. bending stresses, and a variety of less planned loadings resulting from manufacture, transportation, installation and varied (mis) use. Thus elements are often subject to impacts, to localized high stresses often of short duration, and sometimes to 'overstress' (beyond their designed capacity). Tougher materials are better at absorbing the energy of such loads. At the limit, tougher materials are more able to do this locally, through multiple microcracking, rather than allowing a few major cracks to propagate disastrously through the whole building element.

QUALITATIVE IMPROVEMENTS IN TOUGHNESS

So how tough is FC? As mentioned above there is a lack of useful field data for natural fibre concrete. But we can talk qualitatively about how durable fibres will provide some improvement in toughness. This can be thought about by returning to the concept of fracture toughness. Local microcracking is preferable to major crack propagation. Fibres do help by holding the material together, bridging cracks, so inhibiting major crack propagation. And as small cracks propagate, significant amounts of energy can be absorbed by the process of pulling out the fibres bridging the mortar on either side of the crack.

Fibre pull-out is important. Too good a bond of mortar to fibre would result in a more sudden, brittle mode of failure only when the fibres themselves failed in tension. (We see this happening progressively with asbestos cement roofing sheets. It is a fact, generally, that good concretes gain strength slowly over the years. As a result, in asbestos cement sheets the bond of fibre to cement improves with age making the sheets more brittle).

As mentioned earlier and discussed in the next chapter, natural fibres are not very durable in mortar. If this problem could be overcome it might well be worth checking the formulation of FC with respect to fibre length/strength/bond/orientation to improve toughness in situations where it is especially important.

INDICATIVE TEST RESULTS

As mentioned, there are a lack of useful field data for toughness. And there are reservations about the measures that have been used to try to assess toughness of FC. The measures most commonly proposed are impact resistance and the area under the stress/strain curve.

Impact tests

In impact testing - letting a standard weight strike a material at a standard speed - we can get a measure of energy absorbtion which indirectly shows a degree of toughness [5,10,15]. The following set of laboratory figures are impact energies absorbed [10]. (Ignore the units, KJ/m^2, just note the relative figures.)

Element	Energy absorbed (KJ/m^2)
Mortar	3
Short sisal fibre reinforcement	5.5
Long sisal fibre reinforcement	10.5

We cannot conclude from this that short fibre FC is twice as tough as mortar, and that long fibre FC is 3.5 times as tough. We are not measuring toughness but impact resistance (and the test conditions are not known). What we can say is that the results indicate a significant potential for improving mortar toughness by adding fibre, apparently more so for long fibres.

Area under the stress/strain curve

Another indicative measure of toughness sometimes used [9,12] is the area under the stress/strain curve - figure 24. The broad (and loose) argument underlying this is that the area is an indicator of FC's ability to absorb the energy of loading. This is not wrong, but there are several reservations.

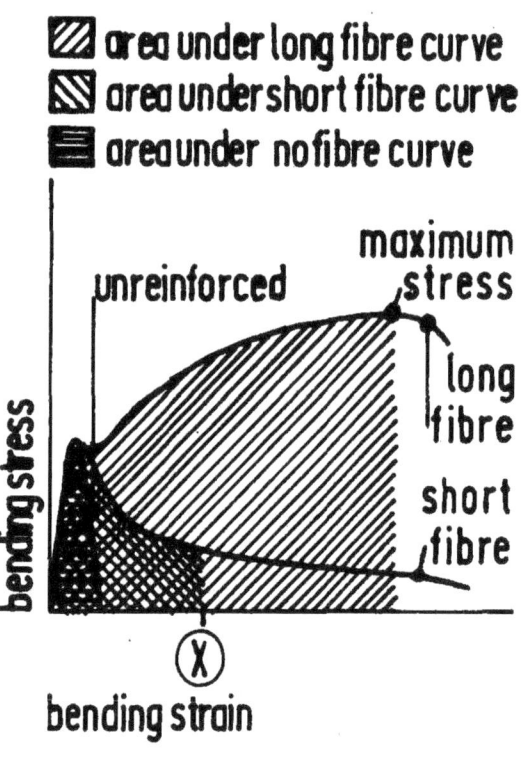

Fig. 24 A diagram that is not as complicated as it first appears. The curves are the same as in figure 21 - stress/strain curves for unreinforced mortar and for FC with long and short fibres. The shaded areas represent areas measured under the respective stress/strain curves as indicators of toughness. As the text explains, there are various (somewhat arbitrary) rules for where to stop measuring the edge of the area along the strain axis. For short fibre FC this could be one per cent strain represented by the point 'X'. For long fibre FC it could be the strain at the point of maximum stress.

First, the area increases significantly as the strain increases so that ultimately FC hangs by a thread (fibre). All this strain capacity is not useful. Much earlier, at less strain, the material becomes unserviceable. So how much of this area should be considered. Sometimes a limit is set at some percentage strain as a cut-off point for measuring the area, but this is clearly somewhat arbitrary.

A second point is that using the area under the bending curve tends to imply that FC of good bending strength is also tough. In fact, it is possible to increase toughness of materials without affecting bending strength, or even reducing it. These two properties - bending strength and toughness - do not necessarily go hand in hand.

Another implication of toughness measurement from bending behaviour is that it focuses on long fibres, which are particularly effective in bending. They may well be more effective for toughness than short fibres because their bond to the mortar is significantly better. But bending strength also favours the aligned orientation of long fibres. Toughness is a more three-dimensional property for which the random orientation of short fibres is more suited. It is likely that the superiority of long fibres over short is being overrated in this measure of toughness.

So what can we say in conclusion? That fibre can bring worthwhile increases in toughness and, on balance, long fibre is likely to be more effective for this than short fibre.

8. DURABILITY OF FIBRE AND FC

A cost-effective method has not yet been found of ensuring long term durability of natural fibres in mortar. The alkalinity of concrete significantly helps protect steel reinforcement from corrosion. Unfortunately, this same alkalinity is destructive of natural fibres. (Biological decomposition does not appear to be a problem, though it could become so if the composite cracked enough to expose the fibres).

FIBRE PRESERVATION

Much work has been done, primarily by Gram[9] on pretreating the fibres to protect them from mortar alkalinity or reducing the alkalinity of the mortar itself. Pretreating fibres has not been successful yet. But a few promising substances have been identified to reduce alkalinity of mortar. Best of these is silica fume, used as a partial cement substitute. It significantly reduces alkalinity of pore water in mortar to the extent that fibre in the laboratory shows no detectable decay in the long term. Silica fume is not widely and cheaply available in non-industrialized countries. However, it is in essence a pozzolana, and there is considerable potential for other pozzolanas, e.g. rice husk ash, to achieve similar results. Further research is required.

LIFESPAN OF FIBRES

The experimental work by Gram was done using laboratory accelerated weathering tests with equipment yet to be calibrated. We do not know how long fibres are being simulated to last. Nor do we have data on actual fibre life within building elements in use. The indications of fibre life are a few years down to a few months.

Factors contributing to this uncertainty include the effects of climate. Test roofs in temperate climates like the UK will be stressed by freeze-thaw cycling, a phenomenon unusual in economically less developed countries. The heat of the tropics will speed up the chemical reactions of fibre decay. So too will frequent wetting and drying since this water movement will help provide the fibres with a fresh supply of alkaline pore water from other parts of the mortar.

DURABILITY OF FC BUILDING ELEMENTS

We can say with some confidence that natural fibres do decay in mortar. What are the implications? At first fibres are likely to be functioning well. Thus they will contribute to manufacturing quality control by improving cohesiveness in the wet state and limiting drying shrinkage cracking when the FC elements are setting. And fibres are still likely to be functioning in the first few stressful weeks of storage and handing, transportation and installation.

FC after fibre decay

We know that FC roof sheets have lasted on roofs for at least eight years[16]. We assume that at least in many of the older sheets significant, perhaps total fibre decay has occurred. These roof sheets remain serviceable in providing shelter.

Inevitably, their strength is that of a well-made mortar. Thus they are brittle, significantly more so than FC with fibres in good condition. The roof sheets need to be treated accordingly. For example people needing to walk on them will require crawl boards to spread their loading on the roof.

Cases of failed roof sheets have occasionally been due to excessive loading, following the assumption that the sheets were stronger than they turned out to be. But in the vast majority of cases, failure is due to poor initial quality control in manufacture and installation. Admittedly, stronger FC (with fibre intact) would stand up better to the rigours of inexpert manufacture and installation.

Some examples have been seen of failure by delamination in roof sheets made by the long fibre method. (Delamination is breaking into thin layers (lamina) like flaky pastry). It is not clear whether this is due to the layers of decayed aligned fibres creating planes of weakness, or if it is due to poor bonding of fibre to mortar during manufacture - figure 4.

SUSTAINED TOUGHNESS

So far we have generalized that fibre decay leads to loss of the strength properties that fibre helps create. However Parry[17], who has more field experience of the technology than anyone, disagrees. On the basis of experienced observation (but not testing) he claims that some of the increased toughness is retained. He notes though that this is "so far without satisfactory scientific explanation".

If he is right in his observations, there could be other causes. For example fibres may not have decayed, or incorporating fibre provides the opportunity for much better manufacturing quality control than could be achieved without it, so that on average FC is a sounder product than mortar alone. These hardly appear good enough reasons to justify his claim.

He is right in saying that there is no satisfactory scientific explanation. Indeed most scientists are against his view. It might be thought that the voids created by fibre decay would help inhibit crack propogation like drilling a hole in a piece of wood or glass to stop a crack spreading. Voids will have some effect, but it is not appropriate to generalize this hypothesis to "holes mean toughness". (Not forgetting that in time many of the holes will fill up with cement hydration products.)

Broadly speaking, the conventional scientific wisdom is that the voids of fibre decay are macrostructural (relatively big) whilst the energy absorbing microcracking occurs on the microstructural (very small) scale, and will not be affected by one or two per cent of large voids. Time and testing will tell.

9. MANUFACTURE

It should by now have become evident that it is worthwhile making and using FC building elements, despite fibre decay. Unreinforced concrete is after all used in a variety of applications: roofing tiles, kerbs, paving slabs, pipes. The loss of strength properties that fibre decay entails in FC is of course a disadvantage. It is to be hoped that research will provide some answers. In this and the next chapter we look at the practicalities of working with FC as it exists today. This is not a complete guide but an indication of key aspects of the technology. Coverage is biased to the most common applications of FC - corrugated roofing sheets and pantiles.

MANUFACTURE

Intermediate technology manufacture is distinguished from other manufacturing by low overheads per workplace, its small scale and decentralised production. Otherwise it has much in common with manufacture anywhere.

Good production engineering - skilled work and supervising the selection of raw materials and their quality control, control of manufacturing and testing of output - is essential. For FC as for all concrete manufacture, water and sand need to be clean, and the sand well graded (though additives can compensate to some extent). Cement needs to be of a high quality and consistent. For FC, fibre needs to be clean and well prepared. Production techniques are not self-evident: they are improving all the time and need to be learned. Testing is needed, such as that described by Parry[16] for roof sheets - bending strength, impact strength, porosity, fibre suitability, consistent shape of products - figure 25. These tests need to be intermediate technology too.

All the above are imperatives, not options, if sound building elements are to be made reliably and consistently. Added to these are the skills of managing, training, distribution and catering for known users or for markets.

AIDS TO MANUFACTURING QUALITY CONTROL

Use of fibre in manufacture has proved its worth even though the fibre decays subsequently.

FC manufacture contrasts well even with some industrialized manufacture without fibre. For example industrialized concrete tiles are made about 12 - 15 mm thick (approx ½ inch) weighing 45-60kg/m^2. FC pantiles can be made half the thickness weighing 21kg/m^2. (However their durability compared with industrialized tiles has yet to be established long term.)

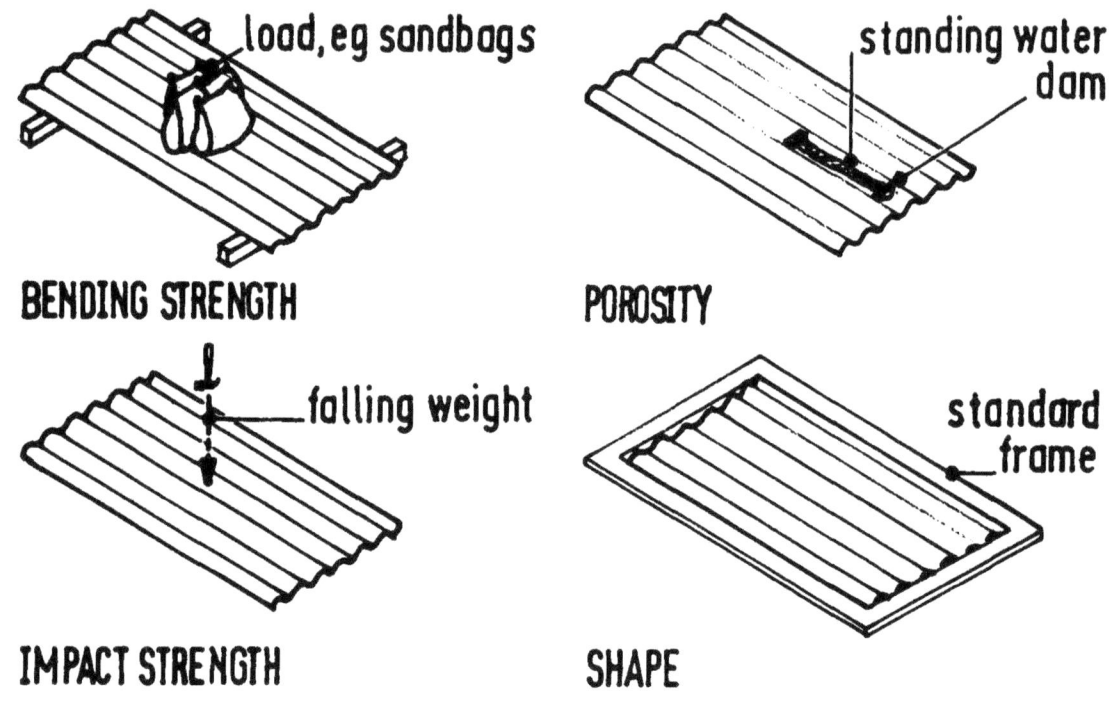

Fig. 25 Examples of low technology testing to maintain quality control: standard loading to check bending strength; standing water for say 24 hours to check if it soaks through as a measure of porosity; a standard weight falling a standard distance to assess impact strength; various standard frames to check aspects of shape such as flatness and squareness. Some of the tests look crude. But the point is not to make exact measurements. Rather it is to demonstrate that minimum quality is achieved and that there is consistency, that is, production quality control.

Several other options set out below have a bearing on manufacturing quality.

Long or short fibres

Until fibre durability problems are solved there is no point in attempting the long fibre method. It is more complex and time consuming and less reliable: it is not yet a well proven manufacturing technology. The short fibre method does prove adequate for providing cohesiveness of the wet mix and drying shrinkage crack control. Long fibre FC also may be subject to delamination (and to extra installation faults mentioned later).

Element size

Where there is a choice, it is desirable to reduce the size of thin FC elements, for example to use a smaller sheet or even a pantile. Smaller elements are easier to trowel to consistent thickness. They should be less prone to serious shrinkage cracking. They are lighter per unit, so should be easier to handle and transport, and will be less prone to overall shape distortion during forming and curing. They will also accommodate irregularities of roof structure better.

Planeness of shape
The need for a cohesive mix to cope with shaping FC, say into corrugated sheets, has been mentioned. Such corrugations are of course structurally necessary to provide better bending strength than a flat sheet. But other structural factors being equal, it is desirable to limit the amount of shaping in order to limit the risk of stretching and cracking in the wet state. From this point of view, roofing sheets with typically six ridges are more difficult to make than pantiles with only one of more gentle curvature - figure 26.

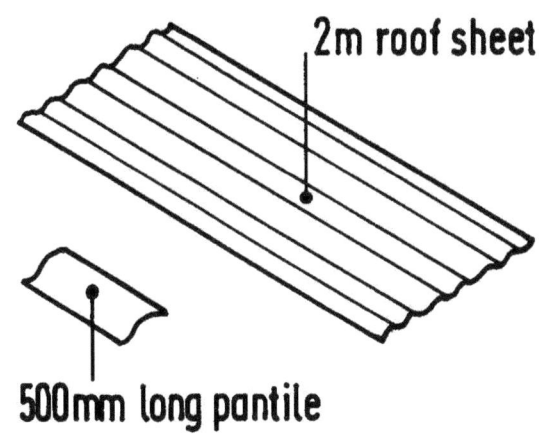

Fig. 26 Pantile and roof sheet drawn to the same scale. The gentler curves of the pantile make it less likely to crack in the wet state (as illustrated in figure 17).

Thickness

A smaller element will carry less load (generally). It can also be trowelled more easily to a consistent thickness. For both of these reasons (and others) it may be possible to reduce the thickness of the element. For example FC roof sheets are typically 8-10mm (approx 3/8 inch) thick weighing 30kg/m^2 covered. Pantiles can be 6mm thick, weighting 21kg/m^2 of roof covered.

Compaction

Compaction of a wet mix consolidates it and drives out air. The usual intermediate technology method is tamping - repeatedly striking the mould or support surface.

Much more effective compaction is achieved by high frequency machine methods. For concrete, this usually involves a mechanical vibrating poker put on the mould or into the mix (which is not appropriate for thin elements). Or it involves electro-mechanical vibration, like the vibrating table illustrated in figure 6. The obvious disadvantages are cost of equipment and the need for a power source - a lorry or car battery in the case of the vibrating table. Hand-powered moderately high frequency vibrators have been tried but nothing yet has proved to be efficient and reliable.

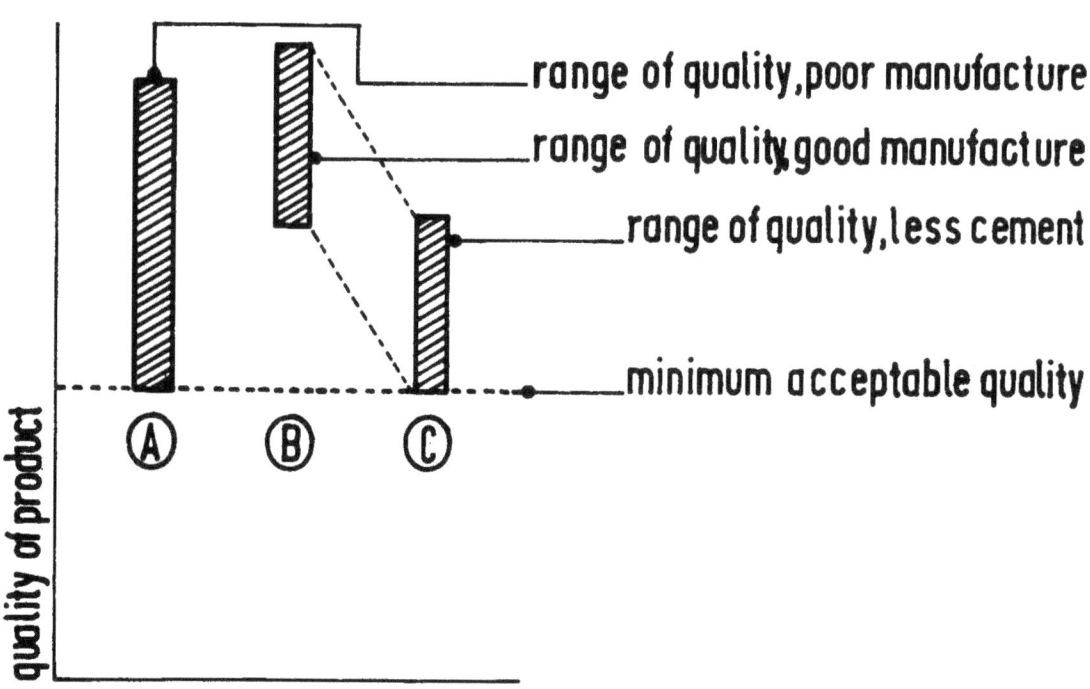

Fig. 27 The tighter the control of production, the smaller and higher the range of product quality will usually be. For example tightening control could change the quality range from 'A' to 'B' in the figure. Once this is achieved, the higher quality could be used/sold. Or quality control could be maintained but the quality values reduced in a controlled way to a still acceptable level - 'C' - by reducing the cement content, so cutting cost.

With good compaction, FC elements are manufactured more consistently and to a higher quality. The greater the compaction, the less the amount of cement is required to produce a quality product (figure 27). This phenomenon is a contributory factor in the cement content differences between roof sheets and pantiles. Roof sheets, hand tamped, have a sand to cement ration of 1:1. Pantiles, electro-mechanically vibrated, have a sand to cement ratio of 3:1. The percentage cement content is reduced from 50 per cent to 25 per cent with the aid of manufacturing quality control.

10. INSTALLATION AND USE

We mentioned earlier the usefulness of bending strength and toughness in FC elements. For example FC roofing is expected to sustain loads such as its own weight, wind pressure, and workmen standing on it. Its toughness is useful for the expected if unmeasured loads of rough handling, nailing, impacts from hailstones and coconuts, etc.

NORMAL BUILDING PRACTICE

More unpredictable but also of great concern to anyone thinking of using FC is the phenomenon known as 'normal building practice'. There are various less polite names for it too. The issue is: whatever you specify, or ask people to do, they tend to do what they normally do. If FC is too special a technology, too different from other technologies, then building with it will almost certainly be a problem.

Normal building practices vary from place to place. For example peple accustomed to roofing with galvanized iron or asbestos cement sheets are likely to find FC sheets much more 'normal' than someone used to thatching. Circumstances alter cases. In contemplating using FC, it is important to check what normal building practice is locally and to establish what needs to be changed for it to be used successfully.

The following sections highlight common situations where FC installation has not been completely normal and has not fitted very readily into existing building practice. They should be treated as examples, giving a feel for possible problems, rather than as a definitive list. Again the bias is corrugated roof sheets and pantiles.

Supporting structures

FC roofing elements should be even in shape. Often roof structures are uneven, especially if made of poles. While galvanized iron sheets can be nailed down, bent and twisted to follow the unevenness, FC is liable to break under such loading. Cracks from poor manufacturing quality control or poor fixing generally occur within the first few weeks of the product's life.

So care must be taken to build roofs as evenly as practical. Sheets should not be fixed down too tightly. Sheets should not bridge across intermediate supports since all the supports are unlikely to lie in the same plane - figure 28. These risks have led Parry[16] to develop half length, that is 1m (40-inch) long sheets. Better still from this point of view are pantiles since they only have to be secured at the head with one fixing rather than multi-point nailing. They are also short, at approximately 500mm (20 inches) long.

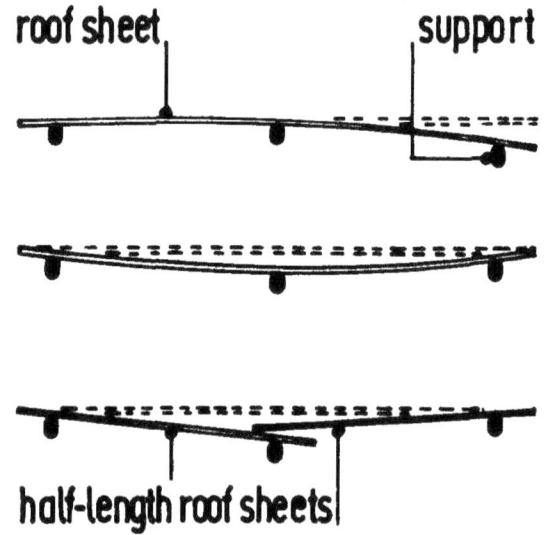

Fig. 28 If supports are not in line, then roof sheets are likely to be bent and risk breaking as in 'A' and 'B'. In 'C' the use of short sheets that span only two supports avoids these risks.

Brittleness

FC roofing becomes brittle with fibre decay. It becomes risky to nail rather than to drill older sheets for fixing during re-use. Where natural fibres have decayed, sheets have about one-quarter the bending strength of asbestos cement sheets. Older sheets will probably be unsafe to walk on, and crawl boards will be needed. If roof structures are likely to move much during their life, such as tied pole structures, then pantiles are preferable because their loose single-point fixing can better adjust to the movement. Where sheets are used, movement can cause the sheets to bind against each other as the roof moves (or if they are badly fitted). In these stressful circumstances, even small movements, due to thermal and moisture movements making the FC change size, could be significant contributors to damage. (Typical moisture and thermal strains for typical climate changes are of the order of 1mm per metre).

Changing loads

Loads may change through the element's predicted life. For example, one very common variation is wind load. As the wind blows across a roof it presses on the near side and causes a suction effect on the other - figure 29. Thus sheets on different sides of the roof are oppositely loaded. And when the wind direction changes the loading pattern is reversed.

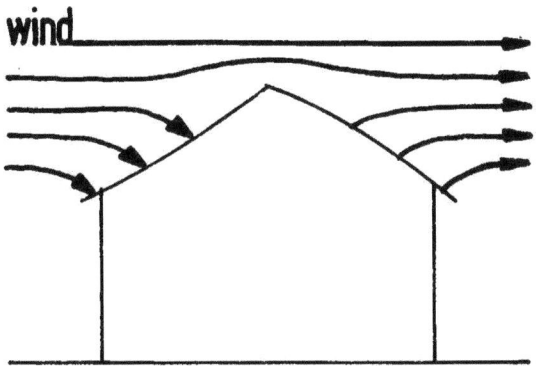

Fig. 29 Wind flow over a building roof. On the exposed (windward) side the roof is pressed down. On the sheltered (leeward) side there are uplift forces. If the wind direction reverses, pressure and uplift forces are reversed. Thus a roof sheet under pressure (windward) will be bent down, subject to compressive stress toward the top surface, tensile stress toward the bottom surface. When the wind direction reverses the surface previously subject to pressure now experiences uplift, and vice versa. Compression on top, tension on the bottom reverses to tension on the top compression on the bottom. Reinforcement on only top (or bottom) would not be appropriate. (Such reinforcement location is common with the long fibre method, though roof sheets could be reinforced throughout by sandwiching more layers of long fibre.)

In general, the changing pattern of loading including any likely changes of use of the building or its parts need to be taken into account at the design stage.

For a definitive treatment of aspects of manufacture, installation and use of sheets and pantiles, see reference 16.

A HAPPIER NOTE

It would be wrong to read this last chapter as a chapter of accidents, as though FC were an inherently risky material. What we have been describing is a new material. And given FC's potential, it is to be hoped it will be a new material in many new places in future. However its particular characteristics have to be understood so that it can be used in an informed way.

11. CONCLUSION

FC is a material with a variety of potential uses, a material that can be manufactured as an intermediate technology now. All the normal care of manufacturing processes needs to be exercised to produce good quality building elements reliably. Consistent quality is essential if the technology is to be a success. Skill is also needed for the installation.

Natural fibres do decay in time, leaving a usable but brittle material. However, fibre is necessary for controlled quality in manufacture. Since manufacturing benefits can be achieved with short or long fibres, there is little point at present in going to the extra effort and greater quality control risks of manufacturing by the more difficult long fibre method. Similarly, in the case of roofing, pantiles have been found easier to manufacture consistently and install effectively than roof sheets. Research is continuing on fibre durability and it is hoped that in future the strength and semi-structural potential provided by natural fibre will be realized and sustained.

In the meantime, natural fibre concrete is providing serviceable products when it is appropriately produced and installed. FC now has a track record over several years in many countries in the form of roofing products, initially as roof sheets and latterly also as pantiles. With care, FC technology offers great promise in providing shelter.

NOTES AND REFEERENCES

1. <u>Fibre cements and fibre concretes</u> by D J Hannant. John Wiley & Sons. 1978.

2. 'Rice husk ash cement and other cementitious materials', by R G Smith, in <u>Appropriate Technology</u>, Vol 11 No 3, December 1984, p8 and 9.

3. <u>FCR: Fibre concrete roofing</u> A field study project carried out by questionnaire on roof sheet installations around the world. Available from the Swiss alternative technology organization SKAT, VarnbitelstraBe 14, CH-9000 St Gallen, Switzerland.

4. <u>The bible</u>, Exodus, Chapter 5, Verse 12.

5. <u>New reinforced concretes: Concrete technology & design Vol 2.</u> R N Swamy (editor). Surrey University Press. 1984.

6. For the more technically minded we can demonstrate the limited order of improvement high modulus fibre would bring, assuming the theory of mixtures. Say the elastic modulus of mortar $E_m=Y$, and that of fibre is 10 times as great, $E_f=10Y$. Assume there is one per cent of fibre and that randomly distributed fibres are only 20 per cent effective (an effective fibre fraction of $0.01 \times 0.2 = 0.002$) and 99 per cent of mortar (a fraction of 0.99). If we call the composite modulus E^C then:

    $$E_c = 0.99E_m + 0.002$$

    $$E_c = 0.99Y + 0.02Y$$

    $$E_c = 1.01Y$$

    Thus even if fibres had 10 times the modulus of the mortar, one per cent of fibre would only add an extra one per cent to strength.

7. In some laboratory tests short fibre length variations have been found to make a small but apparently systematic difference to strength. For example reference 12.

8. Reference 1, p 26.

9. <u>Durability of natural fibres in concrete</u> by Hans-Erik Gram. Report 1/83, Swedish Cement and Concrete Research Institute, S-100 44 Stockholm, Sweden.

10. <u>Natural fibre concrete : SAREC report R2:1984</u> by Hans-Erik Gram, Hakan Persson and Ake Skarendahl. Swedish Agency for Research Cooperation with Developing Countries, S-105 25 Stockholm, Sweden.

11. <u>FICON: Natural fibre reinforced mortar: Report No 2 Technical Characteristics</u> by Malcolm Wilder. Department of Building, Brighton Polytechnic, UK.

12. 'A study of jute fibre reinforced cement composites' by M A Mansur and M A Aziz. In <u>International Journal of Cement Composites and Lightweight Concrete</u> Vol 4 No 2, May 1982, p 75-82.

13. 'The flexural strength of cement-based composites using low modulus (sisal) fibres' by D G Swift and R B L Smith. In <u>Composites</u> Vol 10 No 3, July 1979, p145-8.

14. 'Natural vegetable fibres as reinforcement in cement sheets' by Gladius Lewis and Premalal Mirihagalia. In <u>Magazine of Concrete Research</u> Vol 31 No 10, 7 June 1979.

15. 'Fibre reinforced concrete' by S Sridhara, S Kumar and M A Sinare. In <u>Indian Concrete Journal</u> Vol 45 No 10, October 1971.

16. <u>Fibre Concrete Roofing</u> by John Parry. Intermediate Technology Workshops, Overend Road, Cradley Heath, Warley, West Midlands, B64 7DD, UK. 1985.

17. Reference 16, p25.

www.ingramcontent.com/pod-product-compliance
Ingram Content Group UK Ltd.
Pitfield, Milton Keynes, MK11 3LW, UK
UKHW050227150426
5217IPUK00024B/1677

A technical rationale of how and why natural fibres reinforce cement/concrete, with chapters on the strength properties, durability, manufacture and installation of natural fibre concrete products, fibre-cement, cement, and concrete.

ISBN 978-0-946688-77-7